Steadfast Boats
And Fisher-People

STEADFAST BOATS
AND FISHER-PEOPLE

GLORIA WILSON

Frontispiece: Skipper Eric Smith's 74ft *Rhodella* BCK100 is ready for launching in 1977 from Jones' Buc[...] Shipyard Ltd. She was a typical example of a wooden-hulled transom-sterned seiner trawler built at t[...] time but she was among the first boats to have a fish room cooling plant. Deck shelters and seine ro[...] storage reels were by now standard fittings in the seiner trawler fleet. *Rhodella*'s lines were designed [...] naval architects G.L. Watson & Co. Ltd and she was powered by a Kelvin 500hp diesel engine.

First published 2011

The History Press
The Mill, Brimscombe Port
Stroud, Gloucestershire, GL5 2QG
www.thehistorypress.co.uk

© Gloria Wilson, 2011

The right of Gloria Wilson to be identified as the Author
of this work has been asserted in accordance with the
Copyrights, Designs and Patents Act 1988.

British Library Cataloguing in Publication Data.
A catalogue record for this book is available from the British Library.

ISBN 978 0 7524 5608 9

Typesetting and origination by The History Press
Printed in Great Britain

CONTENTS

ACKNOWLEDGEMENTS

I would like to thank all those kindly and supportive people who provided the information which has enabled me to put this book together. It is impossible to mention them individually because there are so many. My involvement with the fishing communities has been a rewarding and happy experience.

AUTHOR'S NOTES

Almost all the captions mention skippers by name. This should not imply that they were necessarily sole owners of the vessels. Often family members, crewmen, fish-selling companies and others would own shares, and in one or two instances the boats were wholly owned by shore-based interests.

It would be too wordy to mention all owners by name in this book, and could perhaps be intrusive.

Loss of fishing opportunities in distant waters, and high demand for fresh fish, prompted bigger trawler-operating companies to invest in smaller vessels in partnership with the fishermen or as full owners.

Imperial measurements are used in much of the book. All but the newer boats were built in pre-metric days.

Skipper Colin Murray's seiner trawler *Arktos II* KY129 sets out from Aberdeen. Completed in 1979, she was eighteenth in the well-liked series of steel-hulled cruiser-sterned 80-footers from Campbeltown Shipyard and had a Caterpillar 565hp motor.

INTRODUCTION

Steadfast Boats and Fisher-People is the third collection of illustrations which celebrates my deep regard for the fishing communities and their boats. It sets out to define the overall direction in the design of vessels built to the requirements of mostly Scottish fishermen from somewhere around 1965 to the late 1980s, this time span being chosen because I was there, and met the worthy people and their boats.

Because I have taken the photographs the book reflects my own experiences and first-hand research and does not claim to be comprehensive. In the main it concentrates on north-east Scotland because this is the area I knew the best.

I have always been fascinated by the way in which each boat has a unique personality and character. Fishermen regarded their vessels as individuals and discussed their merits and peculiarities in critical detail. It is this strong individual spirit which made the Scottish fishing fleet so lively and colourful and rich in diversity and quality. In fact, these Scottish boats were so likeable that a number were built for customers in Ireland and the north-east of England too.

Many of the pictures in my two earlier books *An Eye on the Coast* and *Kindly Folk and Bonny Boats* reflect my preference for the shapely Scottish wooden-hulled cruiser-sterned flydragging seine netters which were among the most well-respected fishing vessel types built during the twentieth century. Developed by the Scots in the 1920s for catching haddock, whiting, cod, plaice and lemon sole on or close to the seabed, flydragging seine netting was still a principal method of white fish capture in Scotland some fifty years later.

Although the cruiser-sterned wooden-hulled craft must remain my favourites, and I include photographs of some of them, the main core of *Steadfast Boats and Fisher-People* is taken up with vessels of a different character. Introduced in the 1960s, the transom stern marked a significant design change in Scottish wooden-hulled boats, and, onwards from 1970 there was also a big move towards cruiser- and transom-sterned steel vessels in the 40ft to 90ft size range.

My shift in emphasis away from the cruiser-sterned beauties has been prompted by two main factors. Firstly, internet website forums and chatrooms used by fishing boat enthusiasts reveal a growing interest in other vessel types. And a look at these boats, together with my two earlier books, will present a more balanced appraisal of developments in an era which remains strongly within the memory of many people.

In order to illustrate design changes chronologically, the majority of pictures show the boats as they looked when built. Many were later modified and some even quite dramatically altered.

Politically and economically the 1970s were peculiar years, with harmful phases of uncertainty and alarm and downturn. Fisherfolk expressed their unease with port blockades and protest marches. But there was economic growth and the fisheries often prospered enormously.

Changes were taking place. Seine netters fished new grounds further afield in worse weather. Nets made from synthetic fibres yielded massive hauls. Gear-handling machinery was vastly improved by the use of hydraulic winches, the power block and rope storage reels. There were developments in pair trawling, whereby two boats towed a net between them. It caught white fish such as cod and coley on rocky ground, and also colossal amounts of herring and mackerel which swim in shoals in midwater or near the surface. Consequently there were calls for capacious hefty powerful boats with greater sea range and catch carrying capabilities.

Purse seining, whereby a net shaped like an enormous deep bowl is set around a shoal, was introduced to Scotland in the 1960s and caught pelagic species in immense quantities. Drift-net and ring-net fishing, once the chief means of catching herring in Scotland, declined steeply. In the white fish sector there was a move among smaller vessels towards single-boat trawling.

At the start of the 1970s demand for fish was increasing on a huge scale, giving the Scottish fishermen earnings previously looked upon as unattainable. Almost all species were in very keen demand owing to a growing awareness of the nutritional value of fish for human consumption and the general rise in the costs of other protein foods. The Scottish Sea Fisheries Statistical Tables show that the weight and value of all species landed by British vessels at Scottish ports rose from 375,359.8 tonnes valued at £22,559,323 in 1969 to 477,209.8 tonnes worth £64,061,803 in 1974.

In response to the improving fortunes of the fishermen, boatyards were inundated with orders for wooden and steel vessels, with some firms quoting delivery dates several years ahead. One or two new builders entered the fishing boat market and in all some 400 new vessels, the majority in the 30ft to 79.9ft Registered Length size range, joined the Scottish fleet in the years 1969 to 1974 inclusive. This enormous investment reflected the confidence of Scottish skippers and vessel owners in the future of their fisheries and their eagerness to have efficient boats able to work the new fishing techniques to advantage.

Things continued happily until 1974 after which there was a drop in catches, and the growth in gross earnings began to be overtaken by a savage rise in operating costs, largely stemming from the increase in prices of crude oil imposed by the oil-producing countries. And the economic upheaval forced the closure of several boatyards with half-built vessels left on their slipways.

Things recovered, with higher quayside prices and an easing in the rate of inflation, and the value of all species put ashore in Scotland by British boats increased from £59,266,677 in 1975 to a huge £122,223,044 in 1979.

About 190 boats in the 30ft to 79.9ft Registered Length category were produced in Scotland during the years 1976 to 1981 inclusively. The prosperity of the below 80ft sector of the fleet also reflected shortfalls elsewhere. Loss of access to distant water grounds led to the rapid decline of the once prosperous English deep sea trawling ports and also affected Aberdeen.

Early in the 1970s emphasis was placed on increasing the catching power of boats and fishing gear but later there were new considerations. Heavy overheads coupled with catch quotas led to boats keeping fish in the best possible condition to attract high market prices.

Safety of vessels and crews and better living and working conditions also came to the forefront. The Fishing Vessels (Safety Provisions) Rules 1975, and subsequent amendments, laid down requirements for a range of matters including stability, freeboard and hull strength for UK vessels of 39ft 4in Registered Length and over. In 1977 the Fishing Industry Safety Group was established.

The introduction of the transom stern into the seiner trawler fleet was one of the most radical departures in the design of Scottish wooden-hulled vessels since the development of the cruiser stern after the First World War. Until the start of the 1970s the majority of skippers were happy with cruiser-sterned boats. For some forty years they had proved themselves versatile and easy to handle and were such splendid sea boats, but more fishermen were now favouring the transom stern because it afforded more space aft above and below deck, and some felt that the cleaner underwater lines around the stern produced more speed and towing power.

Design and construction details of the transom stern varied from builder to builder, but basically that part of the hull abaft the sternpost ended square instead of being shaped into the sharp ended cruiser stern. The beam of the boat was carried further aft and the run of the buttocks flattened out to provide greater width at the after end of the vessel.

Several Scottish yards had delivered transom-sterned boats to overseas owners in the 1950s. In 1960 Herd & Mackenzie had foreseen the move towards this design feature in Scotland by building the 37ft prawn trawler *Edindoune* BCK142 which had her deckhouse forward. J.&G. Forbes & Co. built Scotland's first transom-sterned seiner trawler *Constellation* FR294 in 1964 for Skipper Joe Buchan, though she too was of forward deckhouse configuration.

Other builders adopted the transom stern during the following five or six years, and, as the 1970s progressed, wooden-hulled boats with this feature began to outnumber those with cruiser sterns. Although some were of stern fishing layout, the majority were of traditional arrangement with deckhouse and cabin aft where there was less motion in rough weather.

Fishermen often discuss the relative merits of wooden and steel boats. Both materials have advantages and drawbacks. Timber has been used throughout boatbuilding history and is enormously strong. Providing that the material was of excellent quality and the building craftsmanship good, a wooden boat can last a long time and several aged sixty years and more were still fishing in the 1970s.

With a handful of exceptions, steel had been used only for larger trawlers and some steam-powered herring drifters. But in the 1950s and early '60s Aberdeen had acquired some twenty-three diesel-driven steel vessels just below 75ft long as economical successors to that port's ageing and decrepit bigger steam-powered North Sea trawlers. Named 'sputniks' after the 1957 Russian Sputnik earth-orbiting satellites, they were cheaper to build and operate and could work with smaller crews and yet use the same trawling gear.

The move towards steel came later elsewhere. Among some fifty boats between 40ft and 90ft built for Scottish owners in 1969 only six or seven were steel. Onwards from 1969 there was a greater acceptance of steel within this size range. Pursing and pair trawling called for vessels near on 90ft long, and so many were built of steel, as good quality large scantlings for big wooden boats became difficult to find.

At the same time, cutbacks in demand for other commercial craft caused shipyards with steel shipbuilding skills to compete for fishing vessel orders. The price of steel boats became more competitive owing to modern welding and prefabrication techniques and the rising cost of prime timber. Providing hull shape and other design features were right, steel boats could be as seaworthy as wooden ones of similar size.

Steel is of a uniform quality, so the components of a boat are welded together to form a strong homogenous single-piece unit. Well-built steel craft do not leak and are able to withstand quite hefty knocks and dents without loss of strength. The material can be fabricated into complex structures, so steel boats can have refinements such as a bulbous bow. Ballast tanks and fish tanks are incorporated easily because the hull forms their outer walls. Steel vessels can be modified or lengthened without being weakened and are able to withstand vibrations produced by bigger engines.

Because steel is consistent in quality the plates and shapes are thinner than the scantlings of a wooden vessel, thereby providing more internal space. The disadvantages of steel were being understood and overcome. Corrosion caused by rust and electrolysis could be counteracted by careful use of correctly treated, shipbuilding quality steel and protective coatings, and the fitting of sacrificial anodes. The magnetic effects of steel caused some wheelhouse instruments to malfunction but this could be remedied. Condensation problems and noise transmission could also be reduced.

Some skippers felt that steel boats were more suitable for working the new fishing methods. Pair trawling was very strenuous and wooden-hulled vessels could sustain nasty damage should they bump into their partner boats. The heavy gear and otter boards used by single-boat demersal trawlers could also be harmful.

Others still preferred wooden-hulled boats and the builders of these had full order books. Scottish wooden-hulled vessels were praised all over the world for their excellent sea-keeping and handling qualities. Although wooden boats were vulnerable to leaks and to attack from wet rot fungus and worm, and timber could be of variable quality with internal weaknesses, these faults could be counteracted by larger scantlings and the careful choice of timber and good construction and treatment. Hulls could be sheathed with steel or plastic in the places where abrasion and damage was likely.

During the early 1970s there was a colossal demand for steel boats in the 70ft to 90ft size range, particularly among the herring trawling and purse seining fleets of north-east Scotland. More than thirty were built for Peterhead alone between 1968 and 1977 inclusively. By 1973 at least fifteen British firms as far apart as Campbeltown, London, Hull and Aberdeen were building steel boats for Scottish owners and several skippers ordered steel vessels from the Continent.

Some steel boats broadly resembled their wooden counterparts in shape, although they did not always have the subtleties of line associated with wooden boats owing to the different working properties of the two materials. Other steel craft departed quite radically in shape because of alternative and often more economical construction techniques.

Two Scottish builders of steel boats became big names, Campbeltown Shipyard Ltd and John Lewis & Sons Ltd. In 1972 Campbeltown delivered the 79ft 11in seiner trawlers *Argosy* INS79 and *Ajax* INS82 to skippers Andrew and William Campbell. Skipper William Campbell collaborated with the builders on their design. One of Scotland's most skilful and highly respected seine net fishermen, he was always willing to look at ways of improving vessel efficiency and catch quality and had been awarded the MBE for his services to the fishing industry. Campbell was a great believer in the superiority of Scottish cruiser-sterned boats from the point of view of seaworthiness and gear handling ability.

Argosy and *Ajax* were of round bilge form with cruiser sterns. Their clean, graceful lines were not unlike those of wooden vessels and they were of traditional seiner trawler layout with deckhouse aft. This well-proven design was combined with advanced steel boatbuilding techniques to produce two vessels which were the forerunners of a further twenty-one cruiser-sterned 80-footers built by Campbeltown during the subsequent twelve years. Many seine net skippers preferred the cruiser stern because it gave a continuous smooth support for the ropes.

Meanwhile, during the 1970s Aberdeen built up an efficient little fleet of some twenty-five sea-kindly 86ft white fish side-trawlers. Bigger than the sputniks, they were able to make weekly trips to Shetland and Orkney and the Scottish west coast. Many were built by John Lewis & Sons and became known as the Spinningdale class. Their round-bilged hulls were not dissimilar to those of steam drifters, with rising floors amidships and fullish at the ends for good sea-keeping, but they had transom sterns for ease of construction. Several 86ft seiner-trawler versions were also produced by Lewis.

Independent firms of naval architects were playing a greater role in the fishing industry. Scotland's fleet of seiner trawlers was enhanced during the 1970s by a series of 86ft vessels designed by Tynedraft Design Ltd of Newcastle upon Tyne. The first was *Shemara* PD78, delivered in 1973 from the John R. Hepworth yard at Paull near Hull to Skipper James Pirie of Peterhead.

Skipper Pirie did much to develop herring pair trawling into the vigorous and prosperous industry which it had become in the early 1970s. Tynedraft carried out the entire design project from preparation of hull lines to the décor in the accommodation, but worked in close co-operation with Skipper Pirie whose ideas and preferences were incorporated throughout.

In all some fifteen 86-footers were built to *Shemara*'s lines, but fittings varied in accordance with their owners' wishes. Almost all were ordered by Peterhead skippers who were replacing their existing boats with larger, more powerful craft able to catch and carry more fish.

Owing to the swing towards pelagic and demersal trawling by vessels of 86ft they were designed chiefly as trawlers, but were also capable of seine net fishing. Of round bilge form with transom stern and raked soft nose stem, they were big boats with high carrying capacity and Registered Length of 79ft 11in, beam of 22ft 6in and moulded depth of 12ft. Boats below 80ft Registered Length were classed as Inshore Fishing Vessels and qualified for more generous financial help towards their building than was available to larger boats.

Compared with wooden-hulled craft of similar age and length the Tynedraft boats had deeper bilges and lower, flatter, less hollow floors, and a greater flare to the bow. Although they were of fuller form they were sufficiently fine forward to make good speed (see the body plan on page 96).

Unity PD209, handed over in 1975 from Cubow Ltd of London to Peterhead skipper John William McLean, was characteristic. Commensurate with the move towards greater flexibility she carried trawl winch, seine winch, net drum, net winch with transport roller, anchor windlass, boom swinger and cargo winch. The semi-ring-main hydraulic system enabled the motors in any one unit to be driven from the variable delivery pump powered from the main engine through a step-up gearbox. Pressure controls adjusted the oil flow to meet the demand of the particular unit in use.

For seining *Unity* carried rope coiler and rope storage bins. Her propulsion system comprised air starting B & W Alpha six-cylinder 660hp 413rpm diesel engine with variable-pitch propeller and nozzle. A number of skippers chose this medium-speed engine for its durability and robust design and low maintenance costs.

Blade angle of a variable-pitch propeller can be adjusted to achieve maximum engine efficiency during free-running and fishing. Consisting of a cylindrical steel ring fitted around the propeller, a nozzle gives increased trawl pulling power. For herring searching *Unity* carried sonar which indicated the depth, distance and bearing of the shoals.

Many of my photographs were taken in Peterhead as I spent a lot of time there. From being a failed and woebegone herring fishing centre in the 1960s with weeds growing on the quaysides, this pink granite Aberdeenshire town became the principal British port for the weight and value of all species of fish brought ashore. This awe-inspiring rise to prominence began in 1970 when most seine net skippers boycotted Aberdeen in protest against high landing charges in that grey granite city. Peterhead built up a mighty infrastructure for the fleet, with major harbour and fish market improvements, and by 1980 almost 400 boats and 2,700 fishermen worked from the port.

In that year, the value of all species landed in Peterhead by British boats reached some £33 million as compared with a miserable £725,700 in 1969. In 1988 the total figure stood at near on £67 million, some £63 million being paid for white fish catches, with haddock, cod and whiting being predominant.

These were heady and prosperous days, but there were tragedies nonetheless. During the 1970s and early 1980s the disappearance of several boats with all hands dismayed the fishing communities.

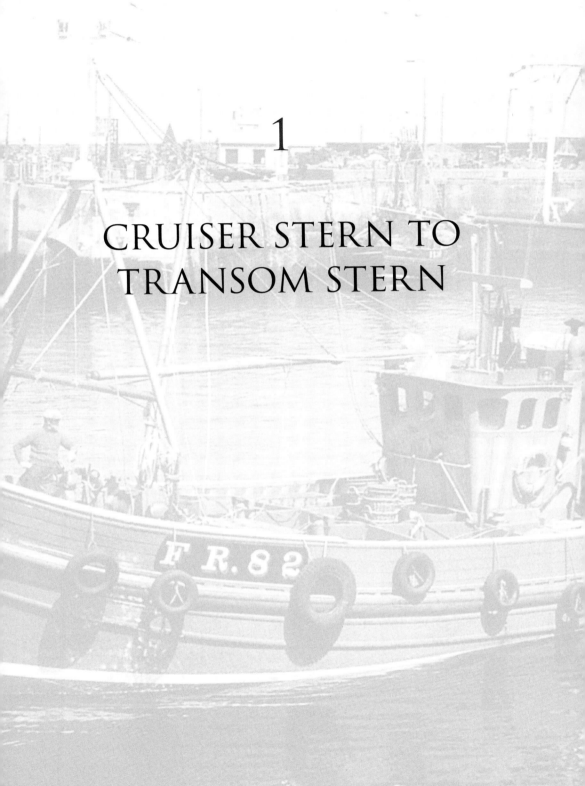

1

CRUISER STERN TO TRANSOM STERN

Cruiser-sterned *Fidelis* KY174
sets off for Whitby after a herring
drifting trip, *c.*1960. The 53-footer
was built in the 1950s by James
N. Miller & Sons Ltd at St Monans
and was powered by a Gardner
114hp diesel engine. Note the
mizzen sail which kept her head
facing the breeze when she was
lying at the drift nets.

Launched late in 1949 from George
Thomson & Son at Buckie, the
cruiser-sterned *Lilt* BCK43 was 66ft long.

Sturdily built 70ft 6in cruiser-sterned seine netter and herring drifter *Manna* BCK 3 was built in 1954 by Alexander Aitken (Boatbuilders) Ltd at Anstruther for Skipper George Murray. She was powered by a Gardner 152hp diesel engine.

Throughout the 1930s Scottish yards had produced dozens of wooden-hulled cruiser-sterned diesel-powered seine net vessels in the 50ft to 60ft size range. *Pentland Firth* BCK99 was built at Lossiemouth in 1932 as *Maggie Fleming* INS255 for that port which, by 1937, owned eighty-one motor boats valued at £150,000.

Seine netting increased at an even greater rate after the Second World War and the fishery was still concentrated largely in the Moray Firth. Measuring 62ft with 18ft 6in beam *Speedwell* INS162 was built for Lossiemouth in 1947 by J.&G. Forbes & Co. at Sandhaven.

Among some fifty boats between 40ft and 90ft delivered to Scottish owners in 1969, all but six or seven were of wooden hull construction. Seiner trawlers *Forthright* KY173 (foreground) and *Steadfast* KY170 were built in 1969 by Richard Irvin & Sons Ltd at Peterhead for Skippers Robert and Alec Gardner. They were 78ft long with Caterpillar 400hp diesel engines.

Wooden-hulled seiners and light trawlers lie alongside the South Harbour fish market in Peterhead, c.1972. *Supreme* PD190 was built in 1957 by Gerrard Brothers at Arbroath.

During the 1960s James Noble (Fraserburgh) Ltd introduced a new series of sturdy, powerful, full-bodied cruiser-sterned trawlers and seiner trawlers which were designed to negotiate turbulent seas. Equipped for nephrops and white fish trawling, the 55ft x 18.2ft 230hp Gardner-powered *Eastern Dawn* FR82 was handed over in 1971 to Skipper Forbes Nicol. She carried a hydraulic winch and power block.

Skipper Robert Smith's 64ft 6in seiner and stern trawler *Argo* FR255 handed over from James Noble in 1966 was designed to her owner's requirements by the Revd Eric Milton, formerly a naval architect. Her transom stern and forward deckhouse provided a roomy working deck. The winch was placed abaft the wheelhouse and the net was worked through an opening in the transom.

Transom-sterned 54ft *Accord* FR11 was built as *Mystic* FR11 by J.&G. Forbes in 1968 for Skipper William Cowe. She was of traditional layout with deckhouse aft and hydraulic winch forward but a towing gantry spanned the stern. Her Kelvin 240hp motor provided good power for trawl towing.

Small stern trawler *Provider* SY24 was delivered from J. Samuel White & Co. (Scotland) Ltd of Cockenzie in 1969 to Stornoway owners. She had a Gardner 200hp engine and hydraulic winch.

Eyemouth Boat Building Co. Ltd handed over the 60ft transom-sterned seiner trawler *Star of Hope* LH260 in 1970 to Skipper Peter Jarron. Built to a new lines plan from naval architect and yard managing director James Evans BSc, she was particularly hefty and roomy for her length. Fittings included a Caterpillar 280hp motor, hydraulic winch and power block, and Beccles rope coiler.

Transom sterns differed from cruiser sterns in their construction. The after frames were set at right angles to the centreline of the boat. This is the 69ft 6in seiner trawler *Morning Star* BF50 being built by Forbes in 1969 for Skipper Harold Napier.

In forming the framework for the cruiser stern, half frames, or cant frames, were set obliquely in way of the curved outrigger. Here the 55ft seiner *Altair* LH418 is under construction in 1963 at Smith & Hutton (Boatbuilders) Ltd at Anstruther.

Altair's bow frames are being faired to accept the planking. Because timber can be sculpted into shape in this way, wooden-hulled boats can have greater subtleties of line than those built of steel.

From the early 1950s onwards, Richard Irvin & Sons Ltd at Peterhead built many large cruiser-sterned boats. Skipper Alec Gardner's hefty 78ft x 22ft seiner trawler *Steadfast* KY170 was launched in 1969 and was equipped with a Caterpillar 400hp 1,225rpm diesel engine.

Transom-sterned 65ft seiner trawler *Green Castle* BF78 was built by Jones' Buckie Shipyard Ltd in 1970 for Skipper George Skene. Her hull lines were designed by naval architects G.L. Watson & Co. Ltd. Watson had supplied lines drawings for motor yachts produced by Jones and this began a close association between the two firms. *Green Castle* was powered by a Caterpillar 345hp engine and was fitted with a power block at the time of her completion.

During the 1970s some seventy-eight fishing boats were built in Scottish yards to G.L. Watson designs. They included the 54ft 260hp Caterpillar-engined transom-sterned scallop dredger and herring trawler *Silver Fern II* CN76 delivered in 1971 from Gerrard Brothers at Arbroath to Skipper Colin Campbell.

Yard Numbers 50 and 51 take shape at Gerrard Brothers to G.L. Watson transom-stern designs, *c.*1971.

Much thought was given to stern trawler design. Built by Forbes in 1970, the 73ft white fish boat *Avenger* BF101 was the result of many months' planning by her skipper, Albert Wiseman, in conjunction with the White Fish Authority's Industrial Development Unit and the builders.

Deck layout and hauling arrangements were designed to reduce manual handling of the gear. Two sloping chutes at the stern were most unusual. When the gear was coming aboard the otter boards automatically slid up the chutes without further manoeuvre. *Avenger* was powered by a Lister 330hp engine and her hydraulically driven split trawl winches were fitted abaft the forward superstructure.

Opposite top: Skipper John Watt's 66ft transom-sterned seiner trawler *Excel* BF110 nears completion in 1971 at J.&G. Forbes. Much of her topside planking was iroko, a hard durable West African timber which was able to withstand a good amount of wear and tear. Her main power unit was a Kelvin 320hp diesel.

Opposite bottom: Excel returns to Peterhead after taking part in the 1975 blockade of Aberdeen. Some 870 boats blockaded Scottish ports when fishermen protested against the unsatisfactory political and economic situation.

Banff boatbuilders John Watt & Sons took over the Macduff Engineering Co. Ltd in the mid-1960s and became the Macduff Boatbuilding & Engineering Co. Ltd. Completed in 1971, Skipper Charles Ewen's 70.4ft seiner trawler *Attain* BF97 was the firm's first transom-sterned boat.

Transom-sterned trawler *Strathgarry* was delivered in 1971 from Gerrard Brothers as SY88 to owners in the Isle of Lewis. Designed by G.L. Watson, the 54-footer was of unusual below deck layout with Caterpillar 250hp engine forward and fish room amidships. She later moved to Peterhead as PD91.

About twenty-five wooden-hulled boats produced by Scottish yards in 1971 had cruiser sterns. The last of the herring ring netters were also built around this time. In 1971 Herd & Mackenzie handed over the attractive 64ft ring netter, scallop dredger and pair trawler *Monadhliath* INS140 to Skipper Alistair Jack. She had a Gardner 230hp engine.

The second transom-sterned vessel from Macduff Boatbuilding was the 55.9ft seiner trawler *Sheigra* UL63 delivered in 1971 to Skipper Archie McCallum. Seen here ready for launching, she was the first boat designed by The Napier Company (Arbroath) Ltd, which was set up in 1969 by naval architect and marine consultant Maurice J. Napier FRINA, MCMS.

In 1972 North Shields seine net skipper Cliff Ellis took delivery of the 70ft *Lindisfarne* BCK147 from Jones' Buckie Shipyard. With lines designed by G.L. Watson, she had a stout 22.5ft beam and was the first new boat built in Scotland to have a Baudouin engine, in this instance a 430hp 1800rpm model with variable-pitch propeller.

Skipper John Cowie's 65ft trawler *Gem* BCK213 prepares for sea trials in 1972. She was built by George Thomson to a James N. Miller design and was powered by a Gardner 230hp engine. Note the net drum aft. Adequate space for fittings of this type was a chief advantage of the transom stern.

Differences between the cruiser stern and the transom stern in wooden-hulled boats are clearly seen here. Built to G.L. Watson lines, the transom-sterned 54ft seiner trawler *Carona* KY179 was delivered from Gerrard Brothers in 1973 to owners in Cellardyke. The unidentified cruiser-sterned craft is prettier and more shapely but has less room aft.

Skipper David Cowie's transom-sterned seiner trawler *Stella Maris* WK142 was launched in 1973 from Jones' Buckie. She was 55ft long with 18ft beam and Gross Tonnage of 24.99 under Scottish Part IV Registry. Skippers of vessels under 25 tons did not require a seagoing 'ticket'. Her lines were from G.L. Watson and she carried a Caterpillar 280hp motor.

Alexanders A177 starts out from Fraserburgh for sea trials. Skippered by Alex Malcolm the 72ft 280hp Kelvin-powered seiner trawler was built in 1972 by J.&G. Forbes. Her superstructure including deckhouse, whaleback, masts and spars and seine derricks, as well as the fish room hatch, were aluminium, being about a third of the weight of steel structures of similar size.

Time-honoured boatbuilders James N. Miller, founded in 1747, remained busy. In 1973 it handed over the 55.2ft seiner trawler *Clonmore* KY340 to Skipper David Boyter.

Macduff Boatbuilding built the 55.2ft x 17.9ft transom-sterned *Antares* BF156 in 1972 for Skipper William G. Watt. She was powered by a Gardner 200hp motor.

J.&G. Forbes delivered the 70ft x 20ft 6in seiner trawler *Volente* PD92 in 1973 to Skipper James Milne Watt. He chose a Baudouin 380hp 1,600rpm engine because it was small enough to be installed in a 70-footer without taking space away from the fish room and yet provided the required power.

Yellow-hulled transom-sterned 65ft x 20ft seiner trawler *Elegant II* BF106 was handed over from Jones' Buckie in 1971 to Skipper William Mair and landed her maiden catch at Peterhead in April. Her lines were designed by G.L. Watson and she was powered by a Kelvin 320hp engine.

Herd & Mackenzie built the 56ft *Lynn Marie* BCK86 in 1973 for Skipper William G. Wilson. Powered by a Volvo Penta 270hp engine, she trawled for white fish and nephrops from Buckie and Lochinver. At the time of her building, 'Herdies' had another dozen or so vessels on order. They were productive times.

Skipper David Fairnie's 66.7ft seiner trawler *Nova Spero* LH142 was delivered in 1973 from Mackay Boat Builders at Arbroath and was powered by a Caterpillar 380hp engine. Much later, in the first decade of the twenty-first century, registered as CN187, and owned by Cornish skipper Shaun Edwards, *Nova Spero* fished with lines for albacore tuna.

Launched from Jones' Buckie in 1974 and designed by G.L. Watson, the 72.3ft seiner trawler *Athena* LK237 fished under Skipper Magnus J. Jamieson and was powered by a Kelvin 450hp motor.

Capacious 84ft cruiser-sterned seiner trawler *Resplendent* PD38 was delivered from Richard Irvin in 1973 to seine net specialist Skipper David John Forman. Her broad 23ft 6in beam gave her exceptional internal space and enabled her to carry some 900 7-stone boxes of fish. She was powered by a Mirrlees Blackstone 495hp 750rpm engine. Seine ropes were stored in bins.

In the background is Skipper David Smith's *Argonaut III* KY337 built in 1969 by Jones' Buckie. Early in the 1970s she was the first seine netter to fit a deck shelter and rope storage reels.

Opposite top: Gutsy and powerful 80ft x 22.75ft cruiser-sterned green-hulled seiner trawler *Achieve* FR100 sets out from Peterhead in 1972 for sea trials. Built by Richard Irvin for Skipper Andrew Buchan, she had a Caterpillar 450hp motor and carried a hydraulically driven winch and power block which were now becoming standard fittings in the seiner trawler fleet.

Opposite bottom: Stout cruiser-sterned 78ft seiner trawler *Calvados* PD35 was handed over from Macduff Boatbuilding in 1972 to young skipper Peter Strachan. Her 565hp Caterpillar engine drove a variable-pitch propeller. She was the first in Scotland to have an Elac LAZ71 echosounder which displayed expanded echoes from selected sections of the water depth between surface and seabed.

Bonny-looking 79.9ft seiner trawler *Sunrise* FR159 came from Forbes in 1974 and was powered by a Grenaa 550hp 500rpm motor with variable-pitch propeller. Skipper John Tait felt that a cruiser stern gave a boat superior handling qualities and greater fuel economy. Deck shelters were gaining favour.

Skipper Philip Morgan's 79ft 6in seiner trawler *Graceful* PD133, delivered from Richard Irvin in 1974, leaves Peterhead for trials. Her Mirrlees Blackstone engine developed 495hp.

Completed in 1975, Skipper James Green's 80ft cruiser-sterned seiner trawler *Replenish* FR199 was the hundredth vessel built by Richard Irvin. Seen here in Peterhead's dry dock, she had a 496hp Mirrlees Blackstone engine with variable-pitch propeller. Unlike most other Scottish yards Irvin did not adopt the transom stern.

Launched from Forbes in 1975, Skipper Alec Tait's 82.2ft x 23.2ft cruiser-sterned seine netter and herring and white fish trawler *Dayspring* FR198 was powered by a Caterpillar 850hp motor with variable-pitch propeller.

In 1975 James N. Miller handed over the 58.85ft seiner trawler *Constant Hope II* KY100 to Skipper William Gay.

Measuring just under 40ft long to allow her to catch white fish with seine nets legally within certain coastal areas, the 153hp Gardner-powered *Serene* PD58 had a range of fittings normally found only on a bigger vessel. These included combined seine and trawl winch, a power block and rope storage bins. James Noble built her in 1974 for Skipper William J. Strachan.

Skipper William Wilson's 65.10ft trawler *Lily Oak IV* BCK94 enters Buckie. Built to G.L. Watson lines and delivered in 1974 from Macduff Boatbuilding, she was powered by a Kelvin 415hp engine. By the early 1970s the number of boats being constructed with transom sterns greatly outnumbered those with cruiser sterns.

Designed by the Napier Co., the 54ft seiner trawler *Marelann* BF201 was built by Smith & Hutton in 1974 for Skipper George Hay. Coming under new ownership in the early 1970s, Smith & Hutton sadly went into receivership and closed later in the decade.

A large number of transom sterns can be seen among these wooden-hulled boats in Aberdeen, *c.*1976. About a third of all white fish landed there in the mid-1970s came from trawlers and seiners under 80ft Registered Length. With lines drawn up by G.L. Watson, the 50ft *Franchise* A87 was delivered from Gerrard Brothers in 1969 to Skipper Jack Reid.

The 45ft 6in stern trawler *Golden Fleece* KY116 was built by James Noble in 1975 for Skipper Ronald Duncan. Powered by a Caterpillar 190hp engine, she fished some three to six hours' motoring time away from Aberdeen.

2

TRANSOM STERNS
PREDOMINATE

Laid out in the 1930s to accommodate steam-powered herring drifters, Peterhead's slipway had four berths for vessels up to 100ft long. Here some fifty years later the 74ft seiner trawler *Astra* INS193 is manoeuvred onto the slip for inspection. Designed by G.L. Watson, she was built in 1976 by George Thomson & Son for Skipper William Fletcher and was equipped with a Kelvin 415hp engine.

Gerrard Brothers handed over the 54ft G.L. Watson-designed seiner trawler *Nimrod III* KY79 in 1976 to brothers Alec and Charles Imrie. She was equipped with a Volvo Penta 270hp 1,800rpm motor. Seine ropes were stored in bins.

The unassuming 54ft 230hp Gardner-powered *Orion* KY352, delivered from Gerrard Brothers in 1977 to Skipper William Scott, was a key vessel in the development of Scottish seiner trawler layout. Her deck shelter extended forward along the port side to the whaleback. This gave a foretaste of the three-quarter-length shelters which were to have a big impact on gear and catch-handling arrangements.

Orion in the making at the Gerrard yard. Designed by G.L. Watson, she complied with The Fishing Vessels (Safety Provisions) Rules 1975.

Large and hefty 87.7ft seiner trawler *Carvida* FR 347 was delivered from Forbes in 1978 to Skipper Andrew James Buchan. She was the first new boat built in Scotland to be fitted with the Marconi Chromascope K echosounder on which echoes from fish and seabed were shown in different colours according to their density. Her 825hp 900rpm Mirrlees Blackstone engine drove a variable-pitch propeller.

Opposite top: Built in 1977 by J.&G. Forbes for Dublin skipper Tom Ferguson, the 86ft x 23ft trawler *Shenick* D594 had fittings of unusual make for a UK-built vessel including an Anglo Belgian Co. 810hp engine, Bopp winches and Renou Dardel variable-pitch propeller. This made good sense because the Dublin firm Fitco Ltd held the agency and provided service facilities for these manufacturers.

Opposite bottom: Skipper Alec Forsyth's 56ft x 18ft 300hp Volvo-powered seiner trawler *Endeavour II* A297, completed in 1978, was similar in lines to a number of sturdy little transom-sterned vessels built by Macduff Boatbuilding in the previous few years. She was designed by the Napier Co. in conjunction with the builders. Seine ropes were stored in bins.

Working from Aberdeen in 1977, these transom-sterned boats were all under 69ft long. They are, from the outside, *Clonmore* KY340, *Constant Hope II* KY100, *Fidelitas* KY274, *Golden Fleece* KY116, *Nimrod III* KY79 and *Crimond II* KY246. *Fidelitas* was built *c.*1971 by James. N. Miller for Skipper David Dunn and *Crimond II* was handed over from James Noble in 1973 to Skipper William Boyter. The other four are mentioned elsewhere.

Scottish boats shelter on the Tyne during the 1977–1978 English sprat fishery. Nearest are *Rose of Sharon* LH250 and *Sharon Rose II* LH317 belonging to the family run business George Moodie & Sons. Launched in 1971 and fishing under Skipper Alistair Moodie, the 60.6ft *Rose of Sharon* was among the first transom-sterned seiner trawlers built by George Thomson & Son. John Moodie skippered the 70ft x 21ft *Sharon Rose II* delivered earlier in 1977 from Macduff Boatbuilding and powered by a Caterpillar 425hp motor.

Seiner trawler *Ocean Herald II* BCK156 sets out from Peterhead, *c.*1980. Built in 1978 to G.L. Watson lines by Jones' Buckie for Skipper Robert Patient, she was 73.1ft long with a 500hp Kelvin engine.

Wooden-hulled cruiser-sterned craft were still the choice of some. One or two lovely ones were completed in 1978 including Skipper Sandy McPherson's sweet 74.2ft seiner trawler *Harmony* INS257, which was the penultimate cruiser-sterned boat from Herd & Mackenzie.

New ideas in deck and superstructure layout were being explored. On board the 74ft white fish trawler *Constellation II* FR295, built by Forbes in 1978 for Skipper Joe Buchan, the casing containing galley and mess-deck was offset to port and the deck shelter extended from the centreline to the starboard rail. The wheelhouse was set on the centreline above casing and shelter.

Constellation II worked her gear over the stern and a net drum was positioned below the shelter. There was ample space abaft the drum to accommodate the heavy bobbin gear once the trawl had been hauled.

Trawler *Scottish Maid* BF317 handed over from Forbes in 1979 to Skipper John Scott was the first wooden-hulled boat in Scotland to be fitted with refrigerated seawater tanks at the time of her building. These were ideal for maintaining the freshness of herring and mackerel. The 86ft 840hp boat was also the first built in a Scottish yard to have a Hedemora engine, chosen for its relatively low fuel consumption.

Skipper John Scott, right, with crew members on board *Scottish Maid*.

Demersal and pelagic pair trawler *Wisteria II* FR263, delivered from James Noble in 1979 to Skipper Anthony Cowe, also had refrigerated seawater tanks.

Her propulsion system was unusual. Two 330hp Kelvin engines drove a single variable-pitch propeller. Skipper Cowe preferred this make and chose the twin installation because Kelvin did not produce an individual engine which would give the required horsepower. The position of the wheelhouse and shelter was broadly similar to that of *Constellation II* and the net drum lay some 20ft forward of the transom.

Measuring 75ft with 21ft 6in beam, *Wisteria II* was designed by the builders with stability calculated by the Napier Co.

Apart from large purse seiners and deep-sea trawlers, the 75.30ft trawler *Valonia* LK376 was the first newly built boat to have a totally enclosed three-quarter-length shelterdeck with doors at its after end. Designed by G.L. Watson and powered by a Kelvin 500hp engine, she was handed over in 1979 from Macduff Boatbuilding to a partnership of five Shetland fishermen headed by Skipper Leslie Tait.

Skipper Pete Pulfrey worked white fish pair trawls from Grimsby with the 80ft 415hp Kelvin-powered *Wendy Pulfrey* BCK198 built in 1979 by Herd & Mackenzie. At one time Pete was a distant-water skipper but having seen the decline in deep-sea fishing he switched to smaller boats.

Skipper Norman Stewart's 74ft shelterdecked seiner trawler *Ardency* INS262 was delivered in 1980 from Jones' of Buckie. Her two Gardner propulsion engines gave a total of 460hp. A single Gardner unit of that capacity was not manufactured but Skipper Stewart wanted this make of engine.

In the early 1980s there was a further move towards greater beam and depth and fullness of line. Skipper George Wiseman's 69.55ft x 22.5ft demersal and pelagic trawler *Honeybourne II* BF359 was of chunky hull form. Designed by the Napier Co. in conjunction with owners and builders, she was constructed in 1980 by Macduff Boatbuilding and equipped with a Kelvin 495hp motor and fish room chilling plant.

The 65ft trawler *Piscean* FR276 delivered in 1980 to Skipper Lionel Mainprize was the sixth boat built by James Noble for this well-known Scarborough fishing family. Her Kelvin 495hp engine drove the propeller housed in a Kort nozzle.

Skipper Lionel Mainprize, in the patterned sweater, with crew members on board *Piscean*. Yorkshire fishermen worked robust trawls over rough ground and liked the Noble-built boats for their powerful trawl-pulling capabilities.

Skipper Robert Mainprize's 68ft white fish trawler *Margaret Jane* FR297 came from James Noble later in 1980. Designed by the builders to her skipper's needs, she was fitted with a 495hp Caterpillar engine turning a two-pitch propeller in a Kort nozzle. Sonar and four echosounders gave her tremendous fish-finding facilities and enabled her to work the net close to seabed obstructions.

White fish trawler *Steadfast II* FR443 ran her sea trials in fresh weather. Three-quarter-length shelterdecks were becoming more popular. A hatch in the starboard side of her shelter enabled the cod end to be emptied onto the main deck. The 74-footer was handed over from J.&G. Forbes in 1980 to Skipper Sandy West and her Caterpillar 565hp engine drove a variable-pitch propeller.

Demersal and pelagic trawler *Star Award* BF407 was built by Herd & Mackenzie in 1980 for Skipper Maurice Slater. Her Kelvin 415hp motor was particularly powerful for a 56-footer to give good towing effort, and her hull shape was modified slightly to accommodate a larger propeller for an even better pull.

Three-quarter-length shelterdecks permitted different configurations of superstructure and gear-handling machinery. On board Skipper John Armstrong's 55ft seiner trawler *Green Pastures* KY165 the wheelhouse was positioned almost amidships atop the shelter, and the galley was offset to port on main deck level. The winch lay beneath the shelter's after end. Built in 1980 by James N. Miller, *Green Pastures* had a Kelvin 415hp engine.

Mackay Boat Builders handed over the stalwart 60.3ft x 19.36ft trawler *Berachah* B322 in 1980 to Portavogie skipper William H. Kyle. Her name refers to a Biblical place, the Valley of Blessing. Equipped for white fish, herring and nephrop trawling, she was powered by a 398hp 1,650rpm Deutz engine.

Having come into general use in the late 1960s, hydraulic power enabled gear-handling machinery to be in any required position. Seiner trawler *Good Design III* KY151, delivered to Skipper John Watson in 1980 from Miller, carried seine rope reels at the stern.

Internos BCK239 sets out from Buckie, *c.*1980. Many older seiners by now had power block and deck shelter. The 69ft cruiser-sterned boat still fished under seine net skipper Sandy Sutherland who had her built in 1966 by George Thomson.

Seiner *Internos* BCK239 lands a nice catch at Peterhead where in the late 1970s a vast fleet of some 300 flydragging seine netters provided about 80 per cent of the white fish landings. These were heady and prosperous days.

At the start of the 1980s numerous long-lived cruiser-sterned boats still fished. Working as a prawn trawler from Pittenweem, the 48ft cruiser-sterned *Inter Nos* KY168 was built by Smith & Hutton as a seiner in 1957 for local owners.

Belonging to her original owners the Baird family well into the 1990s, the sleek and elegant cruiser-sterned 75ft *Star of Peace* PD324 was delivered in 1961 from Richard Irvin. Initially she worked seine nets but later trawled for herring, sprats, shrimps and nephrops.

Of traditional Scottish type and layout, the 59.9ft white fish trawler *Sapphire* PD285 was built in 1981 by J. Hinks & Son of Appledore in North Devon for Skipper William Robertson. With Registered Length below 16.5m, she was designed by the Napier Co. and fitted with a Kelvin 280hp 1,200rpm motor.

Demersal and pelagic trawler *Quiet Waters III* FR353 carried refrigerated seawater tanks. Delivered from Macduff Boatbuilding to Skipper Albert Ritchie in 1981, the particularly stalwart 75-footer had 23ft beam and was designed by the Napier Co. in association with the builders. Her Hedemora 540hp engine drove a variable-pitch propeller.

Built in 1981, Skipper Benjamin Nicol's 74ft x 23ft *Acacia II* BF390 sets out for trials. She was the eighth boat from Macduff Boatbuilding to have a three-quarter-length shelterdeck. With provision for demersal and pelagic trawling she was powered by a 440hp Kelvin engine.

Skipper John Mitchell's 65ft 415hp Kelvin-powered seiner trawler *Beryl* BF411 runs trials in 1982 before being handed over from Macduff Boatbuilding. Built to G.L. Watson lines and equipped with Kelvin 415hp motor, she was the ninth shelterdecker from this builder. Her fish hatch could be raised or lowered by means of a hydraulically powered ram.

There was great sadness when several vessels were lost with all hands. Buckie was grief-stricken when the 67ft seiner *Ocean Monarch* BF144 went missing. She was reported to have been last heard from on 15 December 1979 in a Force 8 gale, increasing to a Force 10 storm, some 70 miles off Shetland. Fish-boxes and empty life-rafts were later found.

Those who died were Skipper Gordon Taylor and crew members John Reid, Walter Thain, John Clark, Barrie Sudding, William Coull and Alan Sutherland.

Ocean Monarch was built in 1971 at Jones' Buckie Shipyard and was a wooden-hulled vessel with cruiser stern.

Buckie suffered again in March 1981 when the 65ft trawler *Celerity* BCK142 was last heard from on passage through the treacherous Pentland Firth in 70mph westerly winds, squalls of sleet and a very rough sea. Skipper Sandy Bruce was lost along with crewmen Richard Clark, John Innes, George Reid, Francis Goodall and William Grant. *Celerity* was a wooden-hulled cruiser-sterned boat delivered in 1972 from George Thomson.

Twin-rig trawlers towed two nets side by side in order to cover a greater width of seabed and catch nephrops and flatfish lying close to the bottom. *Rosebay III* was powered by a Caterpillar 705hp engine and fitted with a fish room chilling plant and three-quarter-length shelterdeck. Note the stout towing gantry spanning the stern.

Opposite: When the author visited Scotland later in the 1980s, boats were even more full-bodied. At Gerrard Brothers, planking was under way on the 74.8ft transom-sterned twin-rig trawler *Rosebay III* PD65, handed over in 1988 to Skipper William Lawson.

Designed by naval architects S.C. McAllister & Co. Ltd, she carried her 22.63ft beam for much of her length and her deep hull form gave her a good grip in the water but she had a reasonably fine waterline entry. Note her double oak frames.

The author visited Skipper Iain Philip's seiner trawler *Tyleana* BF61 in Peterhead in 1987 when the boat had just landed 484 7-stone boxes of fish, chiefly cod and haddock, from a pair trawling trip. Handed over in 1986 from Macduff Boatbuilding, the voluminous 75.72-footer was the last wooden-hulled fishing vessel of her type built in a Scottish yard to have a cruiser stern.

Measuring 65ft 11in with generous 22ft 10in beam, the full-bodied seiner trawler *Opportune* BF19 was built by Macduff Boatbuilding in 1984 for Skipper David Ewen. She was powered by a Kelvin 415hp 1,200rpm engine.

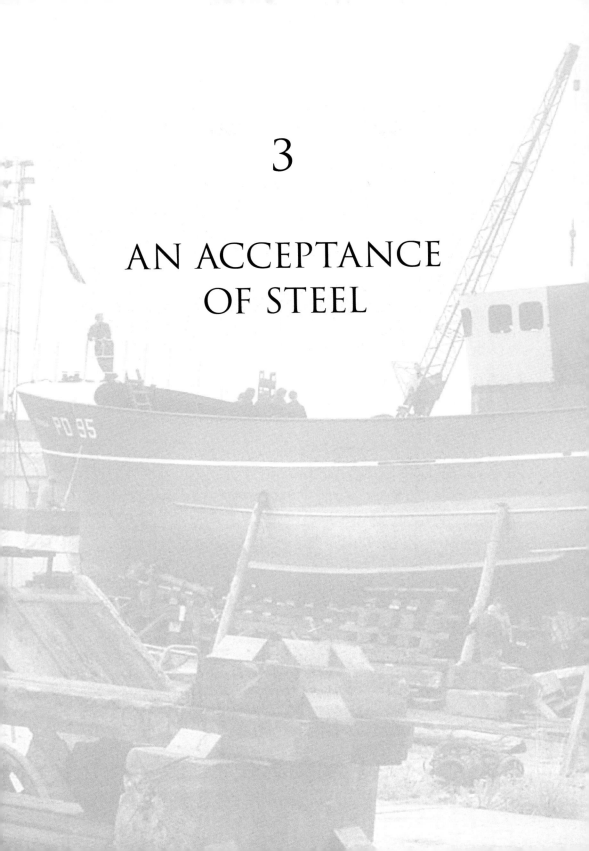

3

AN ACCEPTANCE
OF STEEL

Aberdeen-based shipbuilders John Lewis & Sons Ltd built a dozen 74ft x 19ft 6in sputnik trawlers at its yard in Montrose. Lewis had taken over Montrose Shipyard Ltd in 1959. Completed in 1962, transom-sterned *Lunan* A621, later renamed *Arnisdale* A621, was powered by a Lister Blackstone 330hp motor and carried a belt-driven winch.

Some sputniks found themselves elsewhere. Powered by a Lister Blackstone 264hp engine, the trawler *Confederate* A527 delivered in 1961 from Berwick Shipyard Ltd later joined the Brixham fleet. Between 1956 and 1965 inclusively, the yard produced twenty-two 73ft cruiser-sterned sputniks. Though some were built for seine netting and line fishing, the majority were based as trawlers at Aberdeen.

Cromdale A365, completed in 1969, was an early 86ft Spinningdale class pocket trawler from the Aberdeen yard of John Lewis & Sons, with 20ft 6in beam, Lister Blackstone 495hp engine and a hydraulic winch. The Spinningdales were handy economical successors to the ageing 115ft Aberdeen vessels. They proved themselves to be efficient and sea-kindly high-earners and also had greater sea range than the sputnik trawlers.

Cruiser-sterned seine netter *Challenger* PD104 was produced in 1970 by Richard Dunston (Hessle) Ltd for Skipper Andrew Strachan. She was 79ft 6in long with 22ft beam and powered by a Caterpillar 400hp engine. During the years 1968 to 1977 inclusively, about forty steel boats were built for Peterhead skippers.

Handed over in February 1970, Skipper James Macdonald's 50ft x 16ft prawn trawler and scallop dredger *Crimson Arrow* CN30 was the first boat from Campbeltown Shipyard Ltd which was to become the UK's most productive builder of steel fishing vessels.

Skipper John Horne's 49ft 11½in x 17ft trawler *Steadfast* LH90 took to the water on 2 December 1970 from Campbeltown Shipyard. Note her robust double-chine hull form. Her steering Kort nozzle gave increased towing power but also functioned as a rudder.

Steadfast was roomy for her size with deep freeboard and bulwarks and good below-deck headroom. Propulsion was provided by a Cummins 190hp 1,800rpm engine, and her low-pressure hydraulic trawl winch was mounted forward.

In 1966 purse seining for herring was introduced to Scotland. Delivered in 1968 to Skipper James Lovie from Hugh McLean & Sons Ltd at Renfrew, the 89.7ft transom-sterned *Claben* PD517 was the first steel purse seiner built in Britain for the Scottish fleet. She was powered by a Lister Blackstone 495hp engine.

Peterhead continued its vigorous fleet rebuilding programme with the arrival in 1971 of Skipper Walter Milne's 79ft 9in cruiser-sterned seiner trawler *Faithful II* PD67 from Berwick Shipyard. With 22ft beam she carried a Caterpillar 500hp motor and hydraulic winch and power block and was built primarily for herring pair trawling.

During the late 1960s John Lewis had introduced the stouter MkI version of the Spinningdale class 86ft pocket trawler with 21ft 3in beam. Delivered in 1971 to Skipper Harry Duncan, *Pisces* A193 was second in the series and had a more powerful Mirrlees Blackstone 600hp 706rpm motor. Lewis built the Spinningdales at its Aberdeen premises.

Spinningdale class MkI trawler *Glen Affric* A175 was handed over in 1971 to Skipper William G. Cowie. She too had a Mirrlees Blackstone 600hp engine. Unusually, she fished on a contract basis. Catches were supplied directly to a fish processing factory rather than being sold by auction on the open market.

In working their gear over the stern rather than the side, stern trawlers afforded a higher degree of automation, efficiency and crew safety. Early in 1971 Aberdeen trawler owners Richard Irvin & Sons Ltd took delivery of the 160ft stern trawler *Ben Lui* A166 from John Lewis. Equipped with British Polar 1,650hp engine, she brought back demersal fish chilled in ice, chiefly from Norwegian and Icelandic waters.

At 117ft 9in long the stern trawler *Boston Sea Dart* LT94 was built to fish economically for plaice and cod in the North Sea. Handed over from Hugh McLean & Sons in 1972 to Boston Deep Sea Fisheries Ltd of Lowestoft and skippered by Victor Crisp, she was designed by the White Fish Authority's Industrial Development Unit in conjunction with her owners, with emphasis on safety and automation.

Ben Asdale A328, built in 1972 by Ateliers et Chantiers de la Manche in Dieppe for Irvin, marked a step forward in the design of larger stern trawlers. She was capable of demersal and pelagic fishing anywhere from near to distant water grounds. Measuring 153ft and powered by a British Polar 1,800hp 750rpm motor turning a variable-pitch propeller, she fished under Skipper John Forbes.

The net travels over the stern ramp during *Boston Sea Dart*'s trials off Ayr. Stern trawlers were superior to the traditional side trawlers in that the hazardous task of pulling the bulk of the net by hand over the bulwarks was eliminated.

Built in 1972 for Skipper John Morgan, the 86ft *Seringa* PD95 was the first seiner trawler version of the MkI Spinningdale class of steel vessel from Lewis. She had the same hull form as the trawlers, and Mirrlees Blackstone 600hp motor, but her layout and fittings equipped her for flydragging seine net fishing and herring pair trawling.

Delivered from Lewis in 1972 to Skipper William G. Wilson, the Spinningdale 86ft MkI trawler *W.R. Deeside* A374, later named just *Deeside* BF374, often fished the rough and difficult grounds off the Aberdeenshire coast for only a day or so. Thereby she landed very fresh sole, plaice, haddock, whiting and codling. She too had a Mirrlees Blackstone 600hp engine.

Skipper William Morgan's *Sundari* PD93, built by Lewis in 1972, was the second Spinningdale MkI seiner trawler.

Completed in 1972, Skipper Andrew Campbell's *Argosy* INS79 was the first in the esteemed series of 79ft 11in cruiser-sterned seiner trawlers from Campbeltown Shipyard. They became known as Campbeltown 80s and their lines were not unlike those of wooden-hulled vessels of similar size. *Argosy* had a Caterpillar 480hp motor, hydraulically powered winch and power block and a Beccles rope coiler.

In 1973 Skipper George Murray took delivery of the third Campbeltown 80, the 480hp Caterpillar-powered *Opportune* BCK105. On board his previous boat, the wooden-hulled *Opportune II* BCK60, Skipper Murray had been an early user of a power block and hydraulically driven seine net winch.

Handed over in 1973, Skipper James McPherson's *Cavalier* INS109 was the fourth Campbeltown 80 and was powered by a Mirrlees Blackstone 495hp engine. Seine ropes were stored in bins.

In 1970 seine netters began selling their catches in Peterhead in protest against high landing charges in Aberdeen. Two years later some 150 boats were using Peterhead regularly and this temporary overspill auction hall had been erected to help cope with the influx. Big new harbour and fish market developments were in hand.

Skipper James Pirie was among the first Scots to work the herring pair trawl successfully. In 1973 he took delivery of the 86ft x 22ft 6in 637hp Mirrlees Blackstone-powered *Shemara* PD78 from the John R. Hepworth yard at Paull near Hull. She was the first of some nineteen fishing boats designed by Tynedraft Design Ltd to the requirements of Scottish owners.

Purse seiner and trawler *Pathfinder* BA188 was built in Norway by Kystvaagen Slipp & Batbyggeri A/S in 1973 for former herring ring net skipper Bert Andrew. By 1974 Scotland owned about twenty purse seiners.

Herd & Mackenzie was one of the few Scottish yards to build both steel and wooden-hulled vessels. The 74ft x 22ft cruiser-sterned steel seiner trawler *Ocean Trust* BF307 was built in 1973 for Skipper James Watt. Of round bilge hull form she was the first of a new design from Herdies and was powered by a British Polar 575hp 825rpm engine with variable-pitch propeller.

Late in 1973 Skipper Stanley Morgan took delivery of the 86ft transom-sterned seiner trawler *Summer Dawn* PD64 from Norwegian builder Sigbjorn Iversen Mek. Verksted. Towards the close of the decade she became Peterhead's lone gill netter, catching quality cod on grounds too rocky to be fished by seiners and trawlers.

Salamis PD142 and *Budding Rose* PD84 handed over in 1974 to Skippers Thomas Milne Jnr and James Bruce respectively were the first two 86ft Spinningdale MkII seiner trawlers from John Lewis. They had greater beam and carrying capacity than *Seringa* PD95 and *Sundari* PD93 and more powerful Mirrlees Blackstone 636hp motors.

Multi-purpose 74ft *Mohave* GY309 was equipped for flydragging seining, anchor seining, pair trawling and stern trawling, and could switch readily between these fishing methods. Designed by the White Fish Authority's Industrial Development Unit to the needs of her owners and built in 1974 by Argyll Ship & Boatbuilding Co., she had a Mirrlees Blackstone 495hp engine and fished under Skipper Derek Brown.

Of double chine construction, Skipper Alec Campbell's 56ft stern trawler *Wave Crest* BCK217 was built in 1974 by Herd & Mackenzie. Her Volvo 300hp motor drove through a 4.5/1 reduction gear to the propeller housed in a Kort nozzle. Herdies had first built small steel stern trawlers during the previous decade.

The Spinningdale vessels from John Lewis had demonstrated their catching power. In 1974 a smaller version based on similar lines was introduced with the building of the 75ft 5½in x 20ft 4in seiner trawler *Hesperus* BF219 for Skipper Michael Watt. A neat little craft with agreeable shape and orderly layout, she was powered by a Deutz 450hp 1,500rpm motor, chosen for compactness and light weight.

Mary Croan INS231, built in 1974 for Skipper Tommy Sutherland, was a 75ft version of the resolute series of cruiser-sterned seiner trawlers from Campbeltown Shipyard. With 21ft beam and Part IV Registry under 50 tons she had a Caterpillar 480hp engine. In 1977 she was the top-earning seiner landing at Peterhead.

The flake ice factory in Peterhead provided good viewing points. Steel-built *Seringa* PD95 passes two classic wooden-hulled cruiser sterned seiner trawlers. *Ugievale II* PD105 (right) and *Radiant Way* FR191 were produced in 1968 and 1976 by Richard Irvin.

Just before Christmas 1977, Peterhead pair trawler *Amethyst* PD74 sets out from Plymouth after landing a nice catch of pilchards. Measuring 84ft 9in with 22ft beam and powered by a Mirrlees Blackstone 600hp motor with variable-pitch propeller, she was delivered in 1974 from Southern Shipbuilders (London) Ltd at Faversham to Skipper James Forbes Buchan.

Purse seiners *Comrade* FR122 and *Convallaria V* BF58 were delivered from Maaskant Machinefabrieken in Holland in 1973 and 1974 respectively to Skippers Andrew Tait and Arthur John Ritchie. Measuring just below 90ft, both were powered by Mirrlees Blackstone 750hp engines and had chilled seawater tanks.

Shortly after taking command of his 75ft Campbeltown-built 450hp Caterpillar-engined seiner trawler *Renown* KY257 in 1975, Skipper Albert Smith said she handled well when making for Aberdeen with a Force 10 gale on the beam. Deck shelter and rope reels were fitted at the time of her completion.

Skipper Albert Smith on board his Campbeltown 75-footer *Renown*.

Spinningdale MkII 86ft seiner trawler *Harvest Hope III* PD148 is ready for launch from the John Lewis slipway. Powered by a Mirrlees Blackstone 637hp engine and fitted with rope reels, she was handed over in 1975 to Skipper Peter Stephen.

In January 1975 the 87ft 3in seiner trawler *Sparkling Star* PD137 made her delivery trip across the North Sea in a storm. Built in Holland by K. Hakvoort N.V. for Skipper John C. Buchan, she joined the formidable Big Five team of herring pair trawlers. Her Mirrlees Blackstone 750hp engine turned a variable-pitch propeller.

Skipper John Alec Buchan's 84ft 9in x 22ft cruiser-sterned seiner trawler *Fairweather V* PD157 came from Southern Shipbuilders in 1975 and also joined the Big Five.

James N. Miller & Sons entered the steel fishing boat market with the completion of the 60ft seiner trawler *Sharon Vale* LH304 in 1975 for Moodie Trawlers Ltd. The hull and part of the superstructure were built by McTay Marine Ltd at Bromborough on Merseyside. She was skippered by Douglas Moodie.

In steel boatbuilding during the 1970s there was a trend towards one yard building hulls and another fitting them out. Hull and basic superstructure of Skipper Peter Strachan's 85ft 6in transom-sterned seiner trawler *Stanhope III* PD151 was fabricated by CBS Engineering in Liverpool under contract to the Bute Slip Dock Co. Ltd who completed her. Delivered in 1975, she had a Mirrlees Blackstone 495hp motor.

Five Tynedraft-designed 86ft x 22ft 6in seiner trawlers came from the London firm of Cubow Ltd. All delivered in 1975, they were *Calvados* PD205, *Starlight* PD150, *Unity* PD209, *Day Dawn II* PD136 and *Golden Dawn* PD211.

Skipper Peter Strachan's *Calvados* started life by seine netting from Peterhead. Her Mirrlees Blackstone 850hp 750rpm engine drove a variable-pitch propeller.

Skipper Alex Baird took command of *Starlight*. At one stage all five boats were simultaneously under construction. Two had been launched and were being fitted out while the hulls of the others were in various stages of completion on the stocks. *Starlight* had a B & W Alpha 600hp motor with variable-pitch propeller and nozzle.

Built for Skipper John William McLean, *Unity* was powered by a B & W Alpha 660hp 413rpm engine with variable-pitch propeller and nozzle. A number of skippers chose this medium-speed unit for its robust design and low maintenance costs. Following her completion *Unity* worked the herring pair trawl off the Isle of Man in partnership with *Starlight*.

Skipper McLean (centre) with crew members on board *Unity* in Peterhead. *Unity* was among the first large Scottish boats to pair trawl for cod successfully.

Fishing under Skipper James Tait, *Day Dawn II* initially worked seine nets in the North Sea. Fittings included an Alpha 600hp motor with nozzle and variable-pitch propeller.

Tynedraft-designed 86-footer *Constant Friend* PD83 was delivered from John R. Hepworth in 1975 to Skipper Bruce Thain and firstly worked herring pair trawls around the Isle of Man in partnership with *Starlight* and *Unity*. She had an Alpha 660hp 413rpm engine with variable-pitch propeller in a nozzle.

Delivered later in 1975, Skipper Andrew Cowe's *Golden Dawn* carried the Alpha 600hp engine with variable-pitch propeller and a nozzle. Her deckhouse was slightly shorter than those of the other Cubow seiner trawlers to give greater deck space.

Skipper Andrew Cowe (left) with his crew on board *Golden Dawn* in Peterhead. The 86-footer was the last of the five seiner trawlers from Cubow.

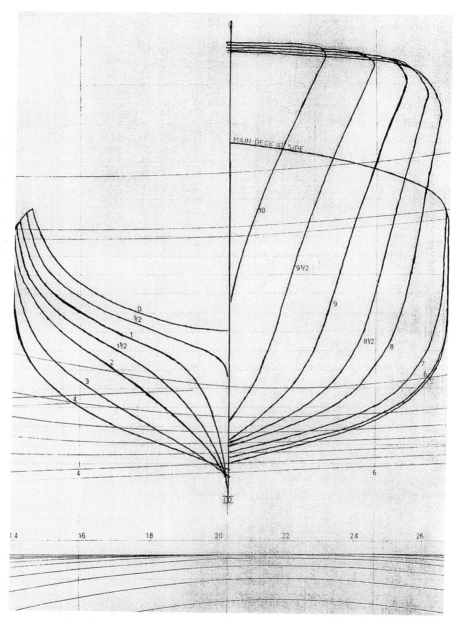

Body plan of the Tynedraft 86ft seiner trawler design. With round bilges, transom stern and raked soft nose stem, they were big boats for their time with high carrying capacity.

Compared with large wooden-hulled boats of similar age, the Tynedraft seiner trawlers had deeper bilges, lower, flatter, less hollow floors and a greater flare to the bow. Although they were of fuller form they were sufficiently fine forward to make good speed. Commensurate with the move towards trawling for demersal and pelagic species by vessels of this size, they were designed chiefly as trawlers but were capable of flydragging seine net fishing. (Drawing courtesy of Tynedraft Design Ltd)

Unity PD209 leaves Peterhead, *c.*1980. In partnership with the Tynedraft 86-footer *Morning Dawn* PD195, she was working the white fish pair trawl, which could be fished over rugged ground.

John R. Hepworth handed over the seiner trawler *Morning Dawn* PD195 in 1975 to Skipper David Morgan. Her Mirrlees Blackstone 637hp engine turned a variable-pitch propeller housed in a nozzle.

On Monday 31 March 1975 eighty boats blockaded Aberdeen as part of the fishermen's protest against the economic and political situation. They were angered by low-priced imports of fish and also called for Britain's territorial waters to be extended. Some 870 vessels blockaded Scottish ports.

Opposite top: Constant Friend PD83 takes delivery of a large Danish-made shrimp trawl in the spring of 1976 in Peterhead. Shrimps were usually caught on the Fladen Ground some 80 to 100 miles from port. The Tynedraft 86-footer in the background is *Morning Dawn* PD195.

Opposite bottom: On Friday 7 January 1977 *Day Dawn II* PD136 landed 120 tons of sprats in North Shields during the winter fishery off north-east England. She was pair trawling in partnership with *Golden Dawn* PD211.

Lying in Peterhead in 1975, the three foreground boats represent the move from cruiser to transom sterns in wooden hulls, and the greater use of steel in the below-80ft Registered Length sector of the Scottish fleet. *Attain* BF97 was Macduff Boatbuilding's first transom-sterned vessel (see page 28). Cruiser-sterned wooden-hulled 75ft *Devotion II* PD127 was built as a seiner, herring drifter and great-line fisher in 1954 by J.&G. Forbes.

Steel *Juneve III* PD215 delivered from John R. Hepworth in 1974 was the second Tynedraft 86ft seiner trawler to enter service and was a member of the Big Five group of herring pair trawlers.

4

PROSPEROUS YET TROUBLED TIMES

This Tynedraft 86ft seiner trawler was caught up in the 1970s boatyard collapses. Originally ordered by Peterhead interests, she was under construction at Ryton Marine (Ship Division) Ltd on Tyneside but in 1973 that firm went bankrupt.

Her hull was taken to Whitby where Intrepid Marine International Ltd planned to fit her out. Following this company's closure she was finally completed in 1975 as *Andrée* BCK160 at B.U.T. Engineers (Grimsby) Ltd. Her original owners had pulled out of the venture. Transferred to British United Trawlers Ltd she fished from Scottish ports under various skippers.

Another Tynedraft 86-footer fell victim to the Ryton Marine failure. Following a sojourn in Whitby her hull also went to B.U.T. Engineers who completed her in 1975 as *Troilus* BCK159. The original Peterhead owners withdrew and she was handed to British United Trawlers.

Troilus later joined the Fraserburgh fleet and was renamed *Ocean Harvest* BF145, fishing under Skipper James McLean.

Purse seiner and trawler *Morning Star* PD122 nears completion in Peterhead. She was in build at the Berwick yard of Intrepid Marine International when in 1975 that firm folded. Skipper James Duncan had the Tynedraft 86-footer towed to Peterhead where he employed local contractors to fit her out.

Completed in 1976 she was equipped with a B & W Alpha 700hp engine with variable-pitch propeller in a nozzle. Note the openings for the bow and stern thrusters.

Morning Star was equipped for pair trawling and flydragging seine net fishing in addition to purse seining. She bore the same name as the wooden-hulled cruiser-sterned *Morning Star* PD234 built for the Duncan family in the 1950s by Thomas Summers & Co. in Fraserburgh.

Skipper Peter Duncan's 80ft cruiser-sterned seiner trawler *Marigold* PD145 was built twice. Southern Shipbuilders went into receivership when she was under construction. Her incomplete hull was cut into pieces and taken by road to Bideford Shipyard 1973 Ltd for rebuild and fitting out. Handed over in 1976, *Marigold* was powered by a B & W Alpha 500hp 400rpm engine with variable-pitch propeller.

In all, Campbeltown Shipyard produced twenty-three cruiser-sterned 80-footers, the majority of them for Scottish owners. *Fear Not* INS197 was built in 1976 for Skipper John McKenzie. Deck shelters and rope reels were by this time standard features on board the seine netters. *Fear Not* had a Caterpillar 565hp engine and fished from Peterhead where in 1976 more than 300 boats landed their catches.

Skipper John McKenzie (right), with crew members on board *Fear Not*.

Spinningdale MkII seiner trawler *Supreme* A476, later INS276, was delivered from Lewis in 1976 to Skipper Innes McPherson. Named after a steam drifter, the 86-footer was powered by a Mirrlees Blackstone 637hp engine.

Lewis used prefabrication construction methods. Hulls were built upside down which permitted down-hand welding. This Spinningdale MkII 86ft seiner trawler was *Helene* PD166 delivered in 1976 to Skipper William Malcolm. Having been bought by the John Wood Group in 1972, the yard now became John Wood Group Shiprepairing Ltd.

By mid-1976 about twenty-four pocket trawlers worked from Aberdeen. *Glen Clova* A607, delivered from Cubow that year to J. Marr (Aberdeen) Ltd, was designed by Tynedraft to her owners' requirements. Fishing under Skipper Kenneth Walker, she was 89.9ft long with 23ft 6in beam and fitted with an Alpha 700hp engine and variable-pitch propeller.

There were more insolvencies. Aberdeen pocket trawlers *Glen Artney* A715 and *Glen Farg* A760 were on the stocks in 1975 when Smith & Hutton ceased to operate. Their 86ft hulls were ready for launch from sub-contractors Tees Marine Services Ltd at Middlesbrough. Designed by Tynedraft for J. Marr, they were completed in 1977 by John Wood Group Shiprepairing.

Spinningdale 75ft 5½in seiner trawler *Castlewood* PD213 was built by Wood Group in 1978 for the Don Fishing Co., but Skipper George Skene and his three brothers owned shares in her. Powered by a Deutz 460hp 1,500rpm motor, she had the same hull lines as *Hesperus* BF219, completed three years earlier.

Campbeltown Shipyard introduced an 85ft class of cruiser-sterned seiner trawler which was designed to be as economical to run as the 80-footers but having greater space and carrying capacity. Third in the series was *Challenger II* PD212, built in 1977 for Skipper Andrew Strachan and powered by a Mirrlees Blackstone 600hp engine.

More than 700 members of the Scottish fishing industry marched through Aberdeen in June 1977 to demonstrate for a 50-mile inclusive territorial zone for UK fishermen. But the EEC Fisheries Commissioner, who was visiting Scotland, said that catch quotas were the answer to fish conservation.

Boxes of white fish go for sale at Peterhead. The value of demersal catches landed at the port from UK boats increased from a meagre £705,938 in 1969 to almost £32 million in 1979.

This notice board on Aberdeen fish quay also expressed everyone's concerns at the time of the protest marches.

Delivered in 1978 from Campbeltown Shipyard to Skipper Stewart Buchan, seiner trawler *Fidelis II* FR319 came seventeenth in the series of well-liked cruiser-sterned 80-footers from the Argyll yard. She was powered by a Mirrlees Blackstone 635hp 750rpm engine.

Owned by Skipper John Reid and the Don Fishing Co. Ltd, the 75ft Aberdeen-based seiner trawler *Merlewood* A270 was commissioned in 1978 from Campbeltown Shipyard. The Don Co. was a subsidiary of trawler owners John Wood Group and *Merlewood* was built as part of the group's move into inshore fishing. Her Deutz 460hp 1,500rpm engine was chosen for its compact design.

At the request of Skipper John W.C. Thomson the 79.7ft cruiser-sterned seiner trawler *St Kilda* INS47, produced by Herd & Mackenzie in 1978, had stability features well in excess of statutory minimum requirements. She was deep and full-bodied with a generous 23ft 7in beam and her superstructure was of low profile to reduce top weight. Fuel and water tanks were also positioned to keep weights low down.

Skipper Thomson fished difficult seining grounds off the Scottish west coast. For working the gear in restricted areas of seabed *St Kilda*'s rope reels were designed to set different numbers of coils of rope for each fishing cycle. The boat had a Caterpillar 725hp engine.

Double-chined 56ft x 18.6ft 365hp Caterpillar-powered stern trawler *Ceol-na-mara* BF352 was built by Herd & Mackenzie in 1979 for Skipper Albert Watt. She was equipped to trawl for nephrops and white fish and pair trawl for sprats. Her winch lay abaft the deckhouse and below the aluminium shelterdeck. The name means 'music of the sea'.

Note *Ceol-na-mara*'s double-chine hull form. In single-chine construction the bottom and sides meet at a sharp angle. The double chine is more complex with additional plating between the two sharp angles and gives a softer turn to the bilge.

Seiner trawler *Resplendent* PD298, handed over to Skipper David John Forman in 1979, was the sixth 85-footer from Campbeltown Shipyard. Her stem had greater rake and the profile of her deckhouse was more streamlined to give an elegant look. Seine ropes travelled aft over the deck shelter top for safety. Fittings included a Mirrlees Blackstone 720hp 750rpm engine.

Generously equipped Spinningdale seiner trawler *Linwood* BF353 was delivered in 1980 from Wood Group to Skipper John Watt, Mate Robert Wilson and the Don Fishing Co. Although only 75ft 6in long, she carried seine rope reels and chilled seawater tanks. Her Deutz propulsion engine developed 600hp at 1,500rpm.

Skipper John Watt (left) and Mate Robert Wilson planned to work pelagic trawls with *Linwood* during the upcoming Minch and Cornish mackerel seasons.

Built to the order of Philip, Joseph and Stanley Buchan and skippered by Charles Brown, the 85ft seiner trawler *Spes Melior* PD397 came from Campbeltown Shipyard in 1979 and had a Mirrlees Blackstone 720hp motor. She was the fourth boat of that name in the Buchan family, the first having been a steam drifter bought in 1948 from Aberdeen. The name means 'better hope'.

Some members of the crew of *Spes Melior*. Skipper Charles Brown stands far left and part owner Stanley Buchan is on the far right.

Advantageous building costs offered by overseas yards attracted some orders for new boats. Skipper William Campbell MBE took delivery of the 85ft 1in x 24ft 1in cruiser-sterned seiner trawler *Acorn* INS237 in 1980 from Johs. Kristensen Skibsbyggeri ApS in Denmark. Designed by the yard in collaboration with Skipper Campbell, she was skippered by his son Andrew. Her Callesen 575hp 425rpm motor with variable-pitch propeller was chosen for fuel economy.

Built in 1980 at the Kristensen yard for Skipper Tommy Sutherland, the seiner trawler *Treasure* INS293 was similar in lines, dimensions and general layout to *Acorn* but had some differences including a Caterpillar 624hp engine. Having joined the huge fleet of white fish boats working out of Peterhead, she landed her maiden trip on 14 January 1981.

Tynedraft 86-footer *Troilus* BCK159 unloads her catch at an awkward bend in the Aberdeen fish quay. In partnership with *Andrée* BCK160 she worked white fish pair trawls for a while, finding a good showing of cod around the Shetland Islands.

Discharging fish at Peterhead early in 1979 in a snowstorm. An abnormally big fleet was based there for the time of year. A poor sprat season, and the closure of the main UK herring grounds as a conservation measure, had forced many boats onto white fish catching.

Major extensions to the harbour and fish market were unveiled in 1976 at Peterhead. Fish sales could now be concentrated in the one market at Greenhill. Later in the decade the market was extended further and could then accommodate a total of 10,000 7-stone boxes of fish laid out in single tiers.

Opposite top: Several boats underwent major conversion to extend their fishing scope and efficiency. *Fairweather V* PD157 was lengthened by some 14ft at the Hakvoort yard in Holland and fitted with refrigerated seawater tanks and a shelterdeck. It was important to land herring and mackerel in first-rate condition.

Opposite bottom: At the start of the 1980s some thirty white fish pair trawling twosomes landed at Peterhead. The method caused the boats to soon look work-frayed. Skipper Ian Smith's 80ft *Alert* FR147 was built in 1974 at Campbeltown Shipyard and in 1980 pair trawled in partnership with *Faithful II* PD67.

In December 1978 the seiner *Acacia Wood* INS205 with her crew of nine failed to come home. Based on her last known position an air search was held in Force 9 winds but nothing was found, although a life-raft later washed ashore in the Shetland Islands. The steel 86-footer was delivered in 1976 from John Lewis. Those lost were Skipper Alec Jack, Alec Jack Snr, Andrew Cadger, Paul Stewart, James Jack, David McLennan, William Stewart, Peter McKenzie and David Morrison.

Opposite top: Boxes of quality line-caught fish at Gourdon. Haddock and cod were the main demersal species landed in Scottish ports.

Opposite bottom: Sadly there were more tragedies and vessels were lost at sea with all hands. In October 1974 the 85ft steel trawler *Trident* PD111 disappeared while on passage from the Clyde to Peterhead. Those who died were acting relief skipper Robert Cordiner and crewmen Tom Thain, Alex Ritchie, George Nicol, James Tait, Alex Summers and Alexander Mair. A Formal Investigation into the cause of her loss was held in Aberdeen in 1975, resulting in a Report of Court which concluded that it was probable that *Trident* took aboard a sea or succession of seas and foundered, the precise cause of the casualty being unascertainable. The Court considered it probable that deficient stability in her design contributed to her foundering.

Following the discovery of her wreck in 2001, a Re-Opened Formal Investigation into her loss was held in 2009–2010 in Aberdeen. This investigation concluded that the most likely cause of *Trident*'s loss was a sudden and catastrophic capsize followed by sinking. The cause of the capsize is attributed to specific seakeeping characteristics of the vessel combined with the prevailing sea conditions at that time. (Crown Copyright fv-trident.org.uk)

Trident was delivered from Bute Slip Dock in 1973, her hull and basic superstructure having been built by Tees Marine Services.

Bought by Lockers Trawlers Ltd of Whitby in 1996, the 55ft trawler and scalloper *Sardius* BF207 fished under Skipper Michael Locker and was powered by a Cummins 400hp motor. She was built by Hepworth Shipyard Ltd in 1988 as *Sardonyx* BF206 for Scottish owners.

Twin-rig trawler *Heather Sprig* BCK181, handed over to Skipper John Smith in 1989, was the first steel boat from Macduff Boatbuilding. Coming below 16.5m Registered Length, she was powered by a Deutz 554hp engine with variable-pitch propeller in a Kort nozzle. Note the semi bulbous bow.

Completed in 1987, Skipper Alec Gardner's 79.92ft seiner trawler *Steadfast III* KY170 was the first fishing boat built by McCrindle Shipbuilding Ltd of Ardrossan. Her cruiser stern was sharper and her floors had more deadrise than those of many steel vessels and were slightly hollow. Her shapely lines were designed by S.C. McAllister & Co. Ltd and she was powered by a Kelvin 650hp 1,350rpm engine.

GLOSSARY AND NOTES

Perhaps there are those who look at these pictures but are new to the fishing industry and vessel design. The following glossary might be of interest to them.

Trawl: Traditionally used for catching demersal fish, trawling employs a funnel-shaped net attached to the boat by wire warps. As it is towed through the water it is kept open by floats and weights and otter boards. Design and working methods were adapted to target other species.

Flydragging Seine Net: This uses a funnel-shaped net attached to the boat by long ropes. The vessel hauls the ropes which converge and herd the fish into the path of the net. In anchor seining, the ropes and net are hauled in by the vessel while she is lying at anchor.

Ring Net: A ring net is used to encircle herring shoals in sheltered waters and is worked by two boats.

Drift Net: Sheets of netting are joined end to end and hung vertically in the water in the path of oncoming fish, which are caught in the mesh by their gills.

Demersal Fish: Fish which live on or close to the seabed.

Pelagic Fish: Species such as herring and mackerel which swim in shoals at intermediate depths between the seabed and surface.

Nephrops: Nephrops norvegicus are popularly known as prawns. The tails are marketed as scampi.

Side Trawlers: Trawlers which work their nets and warps from one side, but this definition became blurred as gear handling techniques evolved.

Pocket Trawlers: Small side trawlers in the 86ft size range, so called to distinguish them from the larger trawlers based at Aberdeen.

Cruiser Stern: A sharp-ended counter which rakes forward at the centreline and has its fullest part at or below the waterline.

Canoe Stern: A sharp-ended counter which rakes aft at the centreline and has its fullest part at the toprail.

Registered Length: The length of the boat measured from the foreside of the rudder stock to the foremost part of the stem.

Gross Tonnage under Scottish Part IV Registry: A calculation based on the length and beam and internal depth of the boat.

Lines Plan: A drawing which describes the curves of a boat's hull. The body plan shows the vertical cross-sections.

Raised or Rising Hollow Floors: A reverse curve in the boat's sloping bottom between the keel and bilges.

Run: The after part of the boat below the water.

Bilge: That part of the hull where the bottom turns upwards to become the side.

Stem: The forward limit of the centreline of the hull, rising from keel to sheerline and normally straight and raked forward in Scottish vessels.

Entry: The forward part of the boat below the water.

Buttocks: That part of the boat where her undersides curve up towards the counter.

Scantlings: The dimensions of the structural components of a vessel.

Bow and Stern Thrusters: Propellers set in tunnels fitted athwartships through the hull and which enable the boat to move sideways.

Bulbous Bow: A forward-projecting bulge at the forward end of the boat below the waterline.

Power Block: A hydraulically powered roller for hauling nets.

Rope Reels and Storage Bins: Storage reels brought a good degree of automation to seine netting and eliminated the dangerous manhandling of ropes. An alternative storage method allowed the ropes to fall through hatches into bins below deck.

The Big Five: Herring pair trawlers could increase their catching capacity by working in teams. Some could be fishing while others were searching for more shoals or taking their catch aboard. Peterhead's 'Big Five' were legendary.

Below 16.5m Registered Length: Replacing the 'below-25 tons' rule and coming into effect in 1980, this enabled new boats without a certificated skipper to be much deeper and beamier and fuller in proportion to their length.

Under 50 Tons Registry: Vessels exceeding 50 gross tons required a certificated second hand in addition to a certificated skipper.

Other titles published by The History Press

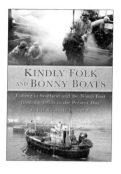

Kindly Folk & Bonny Boats

GLORIA WILSON

A pictorial appreciation of the boats and fishing communities of Scotland and North-East England from the 1950s to the present. From attractive Scottish wooden-hulled craft to recent steel boats, and with many shore scenes including Mallaig herring port, Peterhead harbour reconstruction, fish auctions and fishermen net- and boat-building, this book offers a glimpse into a bygone age. Finally, it considers the work being done to balance fish conservation with profitable fishing, a pressing issue for the fishing industry of the twenty-first century.

978 0 7524 4907 4

An Eye on the Coast

GLORIA WILSON

This collection of photographs and drawings is a personal celebration of fisher people, harbours and boats used along the coast of Scotland and parts of the North East. Often selected for their aesthetic appeal, these pictures show an appreciation of the classic wooden-hulled, cruiser-sterned seine netters and dual-purpose craft. Many of these splendid boats have perished under decommissioning schemes of recent years and so this book could be seen as a lament to a passing era and will undoubtedly evoke many strong memories.

978 0 7524 3853 5

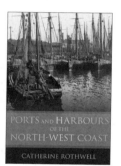

Ports and Harbours of the North-West Coast

CATHERINE ROTHWELL

Who now would link such varied cargoes as slaves, stone, slate, cheese and wine – even shipbuilding itself – with Wardleys, Skippool, Sunderland Point, Silloth, Conishead or Bardsea? More than two centuries ago, shipbuilding was cheaper in the north-west of Britain. This stormy coastline was prized for its craftsmen and many harbours and calling places of refuge developed here. A number have long since disappeared, but their interesting history remains. In this in-depth study, revising and expanding work originally undertaken thirty years ago, Catherine Rothwell looks back at the history of both the 'ghost' ports and such mighty names as Liverpool, Birkenhead, Barrow-in-Furness and Whitehaven.

978 0 7524 5308 8

Sputniks and Spinningdales: A History of Pocket Trawlers

SAM HENDERSON & PETER DRUMMOND

'Sputnik trawler' is a nickname given to two classes of series-built side trawlers. In the mid- to late 1950s and early 1960s, these new and revolutionary boats were intended to replace ageing steam trawlers. The little workhorses had to combat the inevitable prejudice against something ground-breaking and also the torrid economic state of the trawling industry in the 1960s. Inevitably, there were casualties. However, removed from their intended role as mini-side trawlers based at the main trawling ports, the sputniks began to turn in some fine performances for skippers belonging to the inshore ports. Sputniks became successful seine netters, pelagic trawlers and scallop dredgers, their performance often enhanced in later years by extensive rebuilding.

978 0 7524 5452 8

Visit our website and discover thousands of other History Press books.

www.thehistorypress.co.uk

The History Press